Choose The Health Files

Y0-AZF-817

Discovery Channel School Science Collections

© 2001 by Discovery Communications, Inc. All rights reserved under International and Pan-American Copyright Conventions. No part of this book may be reproduced in any form or by any electronic or mechanical means, including information storage devices or systems, without prior written permission from the publisher. For information regarding permission, write to Discovery Channel School, 7700 Wisconsin Avenue, Bethesda, MD 20814. Printed in the USA ISBN: 1-58738-139-7

1 2 3 4 5 6 7 8 9 10 PO 06 05 04 03 02 01

Discovery Communications, Inc., produces high-quality television programming, interactive media, books, films, and consumer products. Discovery Networks, a division of Discovery Communications, Inc., operates and manages Discovery Channel, TLC, Animal Planet, Discovery Health Channel, and Travel Channel.

Writers: Jackie Ball, Kristen Behrens, Michael Burgan, Maxine Dormer, Ruth Greenstein, Monique Peterson, Anna Prokos, Rachel Waugh. **Photographs:** Cover, gameboard, Eyewire/Digital Vision; p. 3, alcohol in body, BSS; p. 6, shoes, PhotoDisc; pp. 8-9, traffic, Corel; p. 8, football helmet, The Everett Collection, biplane, Corel, Ralph Nader, Hulton Getty/Archive Photo; p. 9, bike rider, Corel, helmet, Artville; p. 10, Tarzan, Hulton Getty/Archive Photo, 1950s woman, SuperStock, Inc., bicep, Corel; p. 11, Twiggy, ©Bettmann/CORBIS, David Bowie, Archive Photos, snowboarder, PhotoDisc, woman exercising, Corel; p. 12, cigarette butts, Corel, pack of cigarettes, PhotoDisc, cancerous lung, W. Ober/Visuals Unlimited; p. 15, Jamie-Lynn Sigler, David Allocca/TimePix; p. 16, paint can, PhotoDisc, lightning, Corel, phone, PhotoDisc; pp. 20–21, alcohol in body, BSS; p. 22, Dr. Giedd, NIMH; p. 23, rollerblader, Corel; p. 24, fruit background, Corel; p. 25, pie-eating contest, Corel; pp. 24–25, feet on scale, chewed pencil (both), PhotoDisc; p. 28, D.A.R.E. officer, ©Mary Kate Denny/PhotoEdit; p. 29, D.A.R.E. parade, ©Tony Freeman/Photo Edit. **Illustrations:** pp. 4–5, collage, Elizabeth Brandt. **Acknowledgments:** pp. 14–15, "Losing Control," reprinted courtesy of SEVENTEEN Magazine/Jamie-Lynn Sigler (www.seventeen.com); pp.28–29 "D.A.R.E. to Be Different," excepts reprinted from KEEPING KIDS DRUG FREE, by Glenn Levant. Laurel Glen Publishing ©1998 by Glenn Levant.

Choose

Information, Please!

You're faced with hundreds of choices every day—and the choices you make will affect your health, safety, and well-being. The best way to choose your course in life is to make informed decisions. By knowing the facts about everything from cigarettes and alcohol to food and environmental factors, you can make good decisions. What may be right for you may not be right for your friends or anyone else. It's up to you to make the choices for your life.

In CHOOSE, Discovery Channel gives you the stuff to live a healthy lifestyle. You won't find advice here—but with the information at your fingertips, you'll be able to advise yourself. And that's a healthy start.

The Health Files

Health........4
At-A-Glance The choices you make every day affect your health and safety.

Best Foot Forward 6
Q & A An old pair of sneakers discusses life in the fast—and not always so safe—zone.

Playing It Safe 8
Timeline Where did safety gear come from and why was it invented? Take a ride through time to trace the development of athletic gear and seatbelts.

Hall of Mirrors 10
Picture This Perceptions of a healthy body have changed over time. Investigate some cultural factors behind changing body images.

Up in Smoke 12
Almanac Cigarettes are a drag, especially on your health. Find out what happens to your body when you smoke—and when you quit.

Losing Control 14
Eyewitness Account Many teens think negatively about their own bodies—even those who appear to have a perfect life. See what the reality was for TV star Jamie-Lynn Sigler.

Hazardous to Your Health 16
Scrapbook Danger lurks everywhere—from your home to your school to your neighborhood. You don't have to be an innocent bystander. Take some tips about making choices for a healthy life.

Travel Fever 18
Map Health considerations exist no matter where you go. Get the information you need to make choices for your travel destinations.

Liquor Lowdown 20
Virtual Voyage You're a molecule of alcohol doing some serious damage to a human body. Travel through the major organs to see how you affect them.

Take a ride on an incredible journey. See page 20.

Brain Gain 22
Scientist's Notebook All brains are not the same—that's what scientists found out while studying teenage brains. Their research explains why teens think differently than adults.

Food for Thought........... 24
Amazing But True It's true: You are what you eat. Find out how food increases the energy stores in your body—and the choices you can make for a healthy dose of vitamins.

Mystery of the Misplaced Calories 26
Solve-It-Yourself Mystery What's keeping Graham Baker from losing weight? He's cutting back on calories and eating fat-free foods. What's he doing wrong?

D.A.R.E. to Be Different...... 28
Heroes Glenn Levant dared to challenge drugs. Find out how his organization has helped save thousands of kids from giving in to drugs and alcohol.

Healthy Fun.............. 30
Fun & Fantastic Play a board game for a healthy dose of fun.

Final Project
Safe and Sound 32
Your World, Your Turn Environmental and physical hazards could keep you from living a healthy lifestyle. Investigate your surroundings and make a choice to change them for the better.

THE HEALTH FILES 3

AT-A-GLANCE Health

Decisions, decisions.... You're faced with choices on a daily basis—some difficult and some easy. But every choice you make affects your future, from deciding what to eat for an after-school snack, to where you're going to hang out on a Friday night, to deciding to smoke—or not to smoke. You make choices all the time.

Take the kid on the scooter, at right, for example. He's heading over to a friend's house after school. But he's got a lot of choices to make in the 15 minutes it will take to get there.

He has already made some choices, of course. Before he hopped on his scooter, he chose to put on his helmet. He knows that lots of kids—400,000 to be exact—were rushed to emergency rooms last summer with scooter injuries. But he chose not to wear wrist guards or knee pads—they're not cool.

He's been making other decisions all day. Whether it's cold enough to wear a jacket. What to eat for breakfast and lunch. Now he has to make another choice. On the next block, a bunch of kids he knows call him over. They offer him a cigarette. Without thinking twice, he turns them down. He's tried smoking before—he choked and smelled like an ashtray for the entire day. He doesn't want to go through that again, so he gives them a quick wave and takes off.

He checks out the billboards in the neighborhood: sports cars, alcohol, Web sites—you name it. One sign suggests he drink this brand of liquor. Another gives him a Web address where he can shop for the hottest clothes around. In less than one minute, he's bombarded with signs pressuring him to be a fast-driving, alcohol-drinking fashion statement. And there is still a whole afternoon ahead, in which he and his friend will make other decisions about how to spend their time . . . have fun . . . act responsibly. Any one of those choices, no matter how small, will have consequences.

You probably don't realize it, but a typical day in your life requires making similar difficult choices. Knowing your options and the consequences of your actions will help you make informed, confident choices that will keep you healthy and safe. The choice is yours.

4 DISCOVERY CHANNEL SCHOOL

Best Foot Forward

A pair of old sneakers remembers a life of healthy—and not-so-healthy—activity on a teenager's feet.

Q: We're here at a sporting goods store, looking for an expert on health and fitness. Let's see—

A: Hey, how about us? Down here, on the floor!

Q: Why, it's a pair of sneakers! Old ones, from the way they look—and, um, smell.

A: Yes, some would say we've lived to a "ripe" old age. In fact, that's why we're here. Our owner is buying a new pair.

Q: So you're getting the boot, huh?

A: Yep. Kicked off for the last time. It's kinda tough, because we've gotten attached to this kid. How could we help it? She wore us just about everywhere except in the shower. We'll miss her—even though living wrapped around her feet every day wasn't always easy. Or healthy. Or safe.

Q: What do you mean?

A: She's a nice kid, and she never intended any harm. But she put herself—and us—into some pretty dangerous situations, starting from the very first day she wore us home from the mall. Her brother was driving.

6 DISCOVERY CHANNEL SCHOOL

Q: Did he do something unsafe?

A: No, she did. And actually it was something she didn't do. Motor vehicle accidents kill 10,000 kids every year and injure another 50,000. They're the leading cause of death for kids under 14. But that can all be changed! The risk of death or serious injury can be reduced by at least HALF by doing one simple thing.

Q: Which is?

A: BUCKLE UP! Even if you're in the back seat, which she was, you've got to wear a seat belt. Which she didn't. We all got home safely, but that was just a matter of luck.

Q: What happened after you got home?

A: We leaped into action, big time. Right away she ran outside to shoot some hoops. What a blast! We got a real workout, and so did her body, running around, using large and small muscles, making the heart and lungs work hard so they can stay in tip-top shape. People need 20 or 30 minutes of cardiovascular exercise every day: running, playing sports, fast walking, or riding a bike.

Q: So—your owner was doing something healthy.

A: Yes and no. She was getting good exercise, but our shoelaces kept coming untied. She didn't want to slow down to tie them, so naturally she kept tripping. It wasn't our fault, but we still felt bad. She took one really bad spill and scraped her knee—of course, no kneepads. Finally she went inside for dinner. Whew!

Q: Well, THAT sounds healthy.

A: We thought so, until we noticed her eating habits. Or rather, her non-eating habits. Specifically, her non-fat eating habits.

Q: But aren't fats bad for you?

A: Not all fats. Unsaturated fats, like those in olive oil, are good for you. In fact, your body has to have 'em. Otherwise it breaks down muscle and organ tissue to make energy. And that can make you weak . . . sick . . . even cause health problems that will last a lifetime.

Q: Wow. Sounds like she just couldn't stay away from risks.

A: Right. There was always something. Riding her bike on the wrong side of the road—against traffic instead of with it. Riding without a helmet. Not stopping for stop signs or traffic lights. Not using sunscreen at the beach. You know, a couple of bad sunburns when you're a kid can lead to skin cancer when you're a grown-up. And then there was the time when . . . Wait, here she comes!

Q: That girl? But she looks so nice. So normal. Not at all like someone who lives such a dangerous, risky life.

A: She IS nice and normal. That's the problem. She never realized the risks she was taking.

Q: Well, maybe those fresh new sneakers on her feet will help her change those habits.

A: Afraid not. The feet don't make the choices. All the decisions happen in the brain. We only go where we're told. And as for us, it's time to retire. Or at least, retread.

Activity

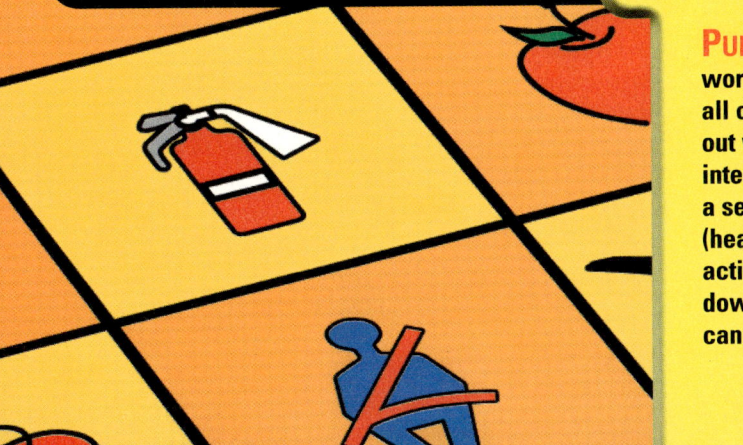

PUMP IT! The harder you work, the harder your heart works. It pumps harder and faster to make sure the muscles all over your body get the blood and oxygen they need. Work out with a friend to measure your pulse at different intensities of exertion. Using a stopwatch or a watch with a second hand, measure your and your friend's pulse rates (heartbeats per minute) after five minutes of each type of activity: sitting down, walking, jogging, climbing up and down stairs. Make a table of your results. What conclusions can you make about the effect of exercise on your pulse rate?

THE HEALTH FILES 7

TIMELINE: PLAYING IT SAFE

🏈 Sports Helmets Through Time

1870s
- Based on soccer and the British game of rugby, football becomes a popular sport at many American colleges. Players wear helmets made of boiled leather.
- Professional baseball catchers begin wearing "birdcage masks" made of an iron frame padded with leather.

1880s
Head injuries among bicycle riders are on the rise as more roads get paved. Some riders wear helmets made of pith, a spongy plant tissue. These offer little protection.

Football helmet made of leather

1900s
- Racing cyclists protect their heads with "hairnets," or strips of soft padding covered with leather. Sweat makes the padding rot.
- In 1905 alone, 18 amateur football players are killed and more than 150 are seriously injured in games. President Theodore Roosevelt meets with sporting gear manufacturers to find ways to improve safety.

1970s
To avoid head injuries, some cyclists wear motorcycle helmets, which are heavy and hot. In 1974 manufacturers create bicycle helmets with Styrofoam™ interiors and stiff exterior shells, held on by nylon chinstraps.

🧍 Seatbelt Story

1880s
E. J. Claghorn receives the first patent for a restraining belt designed to protect passengers riding in horse-drawn carriages.

1910s
A lap belt made of leather is used in a U.S. Army airplane. For the next several decades, seat belts are used mainly on aircraft.

1950s
President Dwight D. Eisenhower's Interstate Highway Act of 1956 funds the construction of America's highway system. The greater number of cars on the road increases the demand for more automobile safety devices.

1960s
Consumer advocate Ralph Nader (above) publishes *Unsafe At Any Speed*, a book that focuses on automobile safety. The U.S. Congress holds hearings on automotive safety. The result is the 1966 National Traffic and Motor Vehicle Safety Act, the first federal safety standards for automobiles.

8 DISCOVERY CHANNEL SCHOOL

Accidents happen in sports and in everyday life.

Thanks to developments in science and industry, we have gear to help us live and play better—and safer. Check out the history of two safety products.

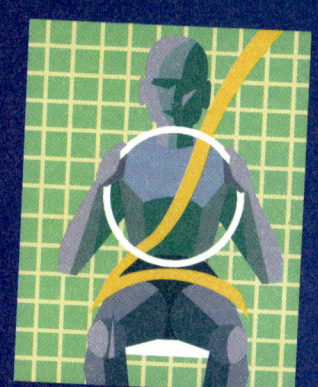

1980s
About 20 different kinds of bike helmets are on the market. A federal bike helmet safety standard is established in 1984; it eliminates the weakest brands.

1990s
Manufacturers introduce a thin, tough plastic for the bike helmet shell and improve designs for a safe, snug fit.

2000
- Air-filled pockets inside football helmets help reduce the number of concussions and other head injuries. Masks attached to the helmets, made of unbreakable plastic coated in rubber, help protect the players' faces.
- New safety features on a baseball catcher's mask include a thick nylon-covered wire exterior and plastic-covered foam padding inside; some also have sun visors and throat protectors.
- Customized to fit a player's head, ice hockey masks (right) are made of fiberglass and kevlar, the material used in bulletproof vests.

1970–80s
- Federal law requires all new cars to have seat belts. Most cars feature separate lap belts and shoulder harnesses. Manufacturers develop the air bag and other passive-restraint systems.
- Many new automobile safety laws are passed: Violators may be punished by law.

1990s
- Studies show that, when used in addition to a seat belt, an air bag may prevent injuries by 68 percent. The 1990 U.S. Highway Act requires that air bags be included in new cars and trucks.
- In 1997 President Bill Clinton signs an executive order that states all federal workers—including military personnel—must fasten their seat belts when driving on government business.

Activity

INVENTION CONVENTION Sure, the inventions on these pages are good, but perhaps you can do better. Challenge your classmates to a competition to come up with a new safety device for one of the following sports:
- Swimming
- Football
- Soccer
- Gymnastics
- Inline skating

Be prepared to defend this device by explaining all safety features. Present ideas to the class to decide which inventions make the sport safer.

THE HEALTH FILES 9

Hall of Mirrors

Trace the ideal body image through changing styles and fashions.

You wish you were taller. You wish you weren't quite so tall. You want to be stronger. You want smaller feet. You want to lose—or gain—a few pounds. In some way you may want to be different from what you are. Most aspects of our appearance depend on genetic factors that can't be changed. One of the few aspects we can control is body weight and shape. You can develop muscles by exercising and lifting weights. You can gain or lose weight by making decisions about what you eat.

Over the years, fashion trends come and go. That's true of body images, too. People often try to reshape themselves according to trends, rather than what is healthiest. Messages everywhere—from advertisements, movies, television, magazines—dictate what's attractive and cool. But what was cool 30 years ago might make you an instant loser today.

Muscle-mania
1930s

Action movies in the 1930s showed audiences the male torso, which was broad-shouldered and thicker in the waist than what we consider ideal today. Here's Johnny Weissmuller in a *Tarzan* movie.

Hourglass Deluxe

The ideal for women in the 1950s was shapely—and impossible for most women to match. Fashions exaggerated this look, which no amount of dieting or exercise could achieve if it wasn't a woman's natural shape.

The Macho Thing
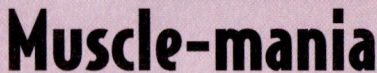
1950s

Bodybuilding for men became big business in the 1950s, thanks to successful advertising from Charles Atlas and Jack LaLanne. The message: If you didn't have muscles, other guys would push you around. "Bulging biceps" was a male star's key accessory.

Thin Is In

Skinny boyish figures were the fashion in the 1960s. The British model Twiggy (left) popularized this trend, which came naturally to the tall and thin. Still, some women went on extreme diets to achieve the look.

1960s

Skinny Stars

1970s

Male ideals changed completely in the 1960s and 1970s. Rock stars like David Bowie (above) set the styles, which included wearing wild clothes and growing long hair.

WORKOUT!

Today people know more about the benefits of exercise. Outdoor activities are popular; inline skating, snowboarding, kayaking, and other such recreation means looking and keeping fit.

2000

Let's Get Physical

The ideal form in the 1980s and 1990s featured firm, toned muscles. Women exercised and dieted to get fit and trim. Even these good habits couldn't make most women look like fashion models and movie stars. Still, they were living a healthful lifestyle.

1980s-90s

THE HEALTH FILES 11

ALMANAC: Up in Smoke

It's no secret that smoking's bad for you. And the numbers don't lie.

Smoke Signals

When people decide to light up a cigarette for the first time, they probably don't realize how hard it will be not to light up again . . . and again . . . and again. Quitting smoking is difficult because, just like cocaine and marijuana, the nicotine in a cigarette is a drug.

Nicotine is a stimulant—and a powerfully addictive one. It increases heart and breathing rates and blood pressure. Plus, nicotine can lead to many life-threatening diseases, including heart and lung sickness, kidney failure, stroke, and cancer. In the United States today, approximately 26.3 million men and 22.7 million women are smokers. Approximately 4.1 million of these smokers are teenagers.

What it Means to Teens

According to the National Household Survey on Drug Abuse, more than 6,000 people under the age of 18 try a cigarette for the first time each day. And more than 3,000 people under the age of 18 become addicted to cigarettes each day. At this rate, 5 million of today's teenagers will eventually die from a disease caused by smoking cigarettes.

Lung cancer isn't the only smoking-related disease that kills. Below is the breakdown of smoking-related deaths in the United States in 1988.

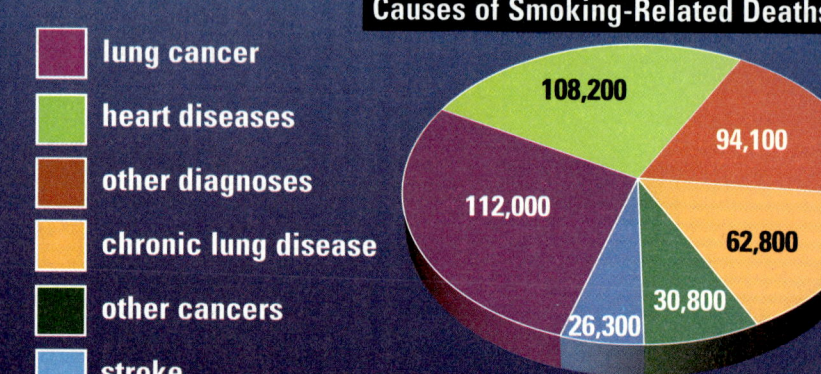

Causes of Smoking-Related Deaths
- lung cancer — 112,000
- heart diseases — 108,200
- other diagnoses — 94,100
- chronic lung disease — 62,800
- other cancers — 30,800
- stroke — 26,300

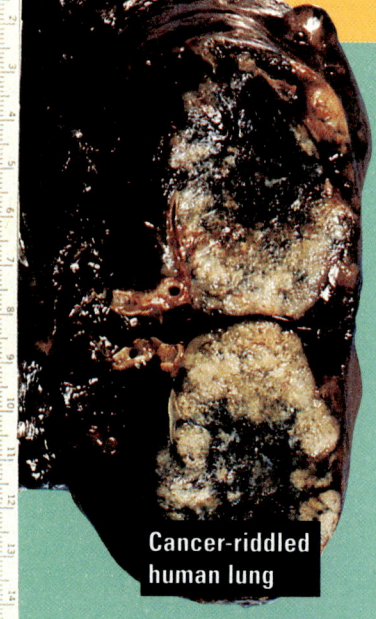

Cancer-riddled human lung

What's Inside?

In addition to addictive nicotine, each cigarette has approximately 4,000 toxins and chemicals—40 of them known to cause cancer in humans and other animals. What kinds of toxins may be rolled into a regular cigarette?

ACETONE:	used to strip paint away from surfaces
HYDROGEN CYANIDE:	poisonous gas
AMMONIA:	strong poisonous chemical used in house cleaners
TOLUENE:	a solvent used to break down heavy-duty products
BUTANE:	the smelly stuff in lighter fuel
NAPTHALENE:	main ingredient in moth balls
METHANOL:	rocket fuel
CARBON MONOXIDE:	exhaust gas from a car
CADMIUM:	used in car batteries

Change Is Good

A smoker's body begins to repair the damage minutes after smoking a cigarette.

AFTER....

20 MINUTES	Blood pressure decreases. Pulse rate drops back to normal.
8 HOURS	Carbon monoxide level in blood drops back to normal. Oxygen level in blood increases to normal. Nicotine level in body is reduced to half.
48 HOURS	Damaged nerve endings all over the body start healing. Lungs start to clear out mucus. Carbon monoxide is gone from the bloodstream.
TWO WEEKS THREE MONTHS	Circulation improves. Walking becomes easier. Lung function increases. Overall energy level increases.
ONE TO NINE MONTHS	Coughing, sinus congestion, fatigue, and shortness of breath decrease.
ONE YEAR	Excess risk of coronary heart disease decreases to half that of a smoker.
5–15 YEARS	Risk of stroke is reduced to that of non-smoker and heart attack risk drops to half that of a smoker. Risk of lung cancer drops to as little as one-half that of smokers.
15 YEARS	Risk of coronary heart disease is now similar to that of non-smokers. Risk of having a stroke is the same as a non-smoker.

Getting Louder

1970
The Surgeon General's warning on cigarette packs:

> Warning: The Surgeon General Has Determined That Cigarette Smoking Is Dangerous to Your Health.

1990s
Studies reveal how smoking is dangerous to your health. The messages, which are rotated on cigarette packs, cartons, and advertisements, cite specific health dangers.

- SURGEON GENERAL'S WARNING: Quitting Smoking Now Greatly Reduces Serious Risks to Your Health.
- SURGEON GENERAL'S WARNING: Smoking Causes Lung Cancer, Heart Disease, Emphysema, And May Complicate Pregnancy.
- SURGEON GENERAL'S WARNING: Quitting Smoking Now Greatly Reduces Serious Risks to Your Health.
- SURGEON GENERAL'S WARNING: Cigarette Smoke Contains Carbon Monoxide.
- SURGEON GENERAL'S WARNING: Smoking Causes Lung Cancer, Heart Disease, Emphysema, And May Complicate Pregnancy.

Supply and Demand

The number of cigarettes purchased in the U.S. has been going down, but total spending for cigarettes has been going up. Why? Because state and federal governments have been steadily raising taxes on cigarettes to raise revenue and discourage people from picking up a bad habit.

NUMBER OF CIGARETTES SMOKED:
- 1986: 594,000,000,000
- 1995: 487,000,000,000

SPENDING FOR THOSE CIGARETTES:
- 1986: $31,800,000,000
- 1995: $45,700,000,000

Activity

TIMETABLES To appreciate the time it takes for your body to heal after quitting smoking, keep a backward diary and match it to the "Change Is Good" timeline on this spread. Write down what you were doing at each interval: 8 hours ago, 48 hours ago, 3 weeks, 2 months, going back as far as you can.

THE HEALTH FILES

Losing Control

Teenager Jamie-Lynn Sigler had an exciting life: good grades, hanging out with friends, a supportive family, and a starring role in a new TV series *The Sopranos*. So why was she depressed—and starving herself?

When her first boyfriend broke up with her, Jamie-Lynn blamed the breakup on her weight. So she decided to go on a diet, not knowing that this first decision was just the beginning of a downward spiral.

I felt like my life was over. I cried for days, moping around. I didn't want to go to dance class, and I didn't even want to go to school. I wondered why he'd broken up with me and began to fear that it was because I wasn't as pretty or as skinny as the other girls we knew, so I put myself on a small diet—just to lose five pounds.

Spinning Off Course

Jamie-Lynn didn't want to keep thinking about her ex-boyfriend. Her solution was to keep busy. She participated in lots of extracurricular activities, even though they took time away from her schoolwork. Her parents worried about her falling grades, but Jamie-Lynn assured them she could handle the load. But three months later, Jamie-Lynn felt her life spinning out of control.

I made what seemed like a great plan. Each morning I'd wake up early to exercise and then choose a menu for the day. Diet and exercise seemed to be the only parts of my life I had complete control over.

I was losing weight and getting tons of compliments from friends and family, which pushed me even further. I severely restricted my calories and woke up earlier and earlier to get extra hours of exercise in. If I didn't work out, I couldn't think about anything else until I did. I began to actually fear food and crave exercise. I was completely miserable, but strangely, I felt I had to do it.

My grades were slipping, and my friends started asking me why I brown-bagged my lunch all the time and never wanted to go out to eat with them. I made excuses, telling them I didn't like the cafeteria food and that I needed to stay behind and catch up on work during lunch period. I didn't go out on weekends anymore—that was when I caught up on my sleep.

Reality Check

Her parents talked about Jamie-Lynn's behavior and weight loss. But she chose to ignore them and even avoided looking at herself in the mirror. People made comments about her weight, but Jamie-Lynn simply brushed these off. Kids in school made fun of her. Reality hit when Jamie-Lynn finally looked at her reflection. She couldn't believe what she saw in the mirror: hollow sunken eyes, ringed with large dark circles, and a sliver of a body.

My bones stuck out, and some ribs were visible . . . I had lost my breasts, my legs were tiny and I had grown hair all over my body [a common characteristic of people who are underweight]. I felt like a ghost.

Jamie-Lynn was so depressed by what she saw that she began to consider extremes to end the pain.

I kept thinking that if I ended my life, this all would go away. That suicide had even crossed my mind scared me more than

14 DISCOVERY CHANNEL SCHOOL

anything had yet, and I began to cry. I told my parents that I didn't want to live anymore, that I couldn't handle it. Finally, I blurted out, "I have an eating disorder. Please get me help!" It was so difficult, but it was the best feeling in the world to finally say it. They admitted they had secretly suspected this all along, but, like myself, had never truly believed that I could have an eating disorder. Then my parents told me they couldn't be prouder that I'd told them and that they were going to help me get better.

Jamie-Lynn and her parents chose to get help—counseling for her depression and medical advice for her undernourished body. She agreed to see a psychiatrist and a nutritionist twice a week. It took her a month to get used to the fact that she needed to gain weight. "Saying 'I gained a pound!' isn't always praised in our society," she says, "but at that time it was the best news I could give anyone."

Back On Track

When it was time to film *The Sopranos* in June, the people on the set were shocked to see Jamie-Lynn looking so thin. They were worried she couldn't handle the long hours of filming. "In short, if I didn't gain the weight back, I would be replaced," she says. Between takes Jamie-Lynn snacked to put on some weight. And she kept consulting her nutritionist and therapist.

Today, my life is pretty much back to normal. I have returned to my regular clothing size, and instead of exercising every day for hours at a time, I go to the gym about three or four times a week... just to be healthy. No foods are off-limits. I make a conscious effort to eat the right foods for energy and good health, but I will not diet. I find myself enjoying life even more now that food isn't always on my brain.

Jamie-Lynn caught herself before it was too late. But that doesn't always happen. Many kids feel they have too much pressure to look good, be smart, have lots of friends, and always succeed. Kids may not know that when they're faced with a stressful situation, it's okay to ask for help—in fact, it's necessary. Some problems are just too big for one person to handle, but other people—a friend, parent, or professional—are there to help. The first step is making the decision to talk to someone.

> "I have an eating disorder. Please get me help!" It was so difficult, but it was the best feeling in the world to finally say it.

Activity

YOU DECIDE Jamie-Lynn's story is one of choices, both good and bad. Usually it's not just one big decision that leads to good or bad health, but a series of smaller ones. Go through Jamie-Lynn's account and identify each case where she made a good decision and a bad one (she mentions at least 15 choices she made). Create a diagram or flowchart to show how one decision led to another. What decisions were behind crucial turning points? Did any choices "cancel out" previous decisions? Then imagine her story if she had done the opposite in each case.

SCRAPBOOK: HAZARDOUS to Your HEALTH

Sights, sounds, and smells surround you wherever you go. Some of them are harmless, even pleasant. Others might present hidden dangers. Read up on what's around you.

GET THE LEAD OUT

Lead is a natural element, but if it gets inside the human body, it causes health problems. High levels of lead cause brain damage, organ dysfunction, and birth defects. Lead is almost everywhere: in the paint in playground equipment and walls, and in many water pipes. Here are some choices you can make to keep safe from lead.

Stay Dust Free Dust can be loaded with lead. Wipe down dusty furniture and floors with water and detergents.

Take 'Em Off Dust and soil that may contain lead can get onto your shoes—and into your carpet. Wipe your shoes on an outdoor doormat, and keep your shoes in a hall closet.

Don't Peel Peeling paint is dangerous. Older paint usually contains lead. Sanding, burning, or scraping lead paint will get it on your hands, in the air, and into your lungs.

Eat Up Vitamins can protect you from lead. Eat plenty of calcium-rich foods, like milk and yogurt, and foods packed with iron, like chicken and dried fruit.

Household Hints

✓ Replace batteries in smoke and fire alarms in your home every six months, and make sure the equipment works.

✓ Never answer or use the phone while you're in the bathtub or shower. Electricity makes a phone ring. If you touch the phone at the exact moment it starts ringing, you could be electrocuted.

✓ If you see lightning, get off the phone. Lightning generates electricity, which can be attracted to a telephone line—and get to you while you're talking on the phone. Stop using your computer if there's lightning in the air, especially if the computer is linked to a phone line.

✓ Most kitchens are crawling with unhealthful bacteria. Handle meat and poultry with care. Scrub your hands, countertops, and other areas that came in contact with the food with warm, soapy water to kill off the bacteria.

If you see lightning, get off the phone.

Secondhand Smoke

If people smoke around you, you become a smoker, too—and that increases your chances of developing a smoke-related illness. In 1996 the Centers for Disease Control and Prevention found that 9 out of 10 nonsmoking Americans are exposed to secondhand smoke. That results in approximately 2 percent, or 3,000, lung cancer deaths each year. When you breathe secondhand smoke, your heart rate and blood pressure go up, and carbon monoxide levels in your blood increase immediately.

To cut back your chances of inhaling secondhand smoke, ask your parents and visitors not to smoke in the house. It may be hard to persuade them, so if they continue to smoke, get out of the room, open windows and flip on some fans. That will help get the air circulating, sending smoke out of your way.

Heading out to lunch with your friends? Ask if you can all sit in the nonsmoking section if the restaurant offers one. If that's not available, sit in the most well-ventilated area—or grab some lunch and eat it outside.

No Cool Camel

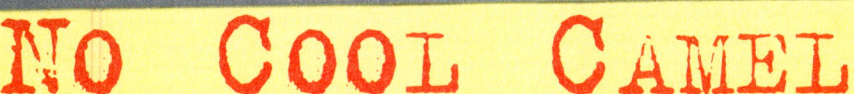

IN 1998 A REPORT RELEASED to the U.S. Congress proved that R. J. Reynolds, the maker of Camel and Winston cigarettes, was targeting underage smokers with ads featuring the cool and slick Joe Camel and his group of leather-clad pals.

A CONFIDENTIAL MEMO in the company stated the ad campaign would use peer acceptance to attract current and potential young smokers. This cigarette company was punished for targeting minors, and other companies received strict warnings from the U.S. Federal Drug Administration (FDA).

NOW TOBACCO COMPANIES cannot use cartoon characters to advertise their cigarettes, nor can they sponsor sporting events. In addition, the companies must spend more than $300 billion over the next 25 years on antismoking campaigns aimed at youth.

THE HEALTH FILES

MAP

TRAVEL FEVER

When you travel, you may have to make some difficult choices.

The World Health Organization, the Centers for Disease Control and Prevention, and other international organizations alert travelers to health and safety concerns. Before you pack your suitcase, pack your brain with these facts.

Map Key
- ❄ Frostbite
- 🫁 Tuberculosis
- 🦟 Malaria
- ⛰ Altitude Sickness
- ☀ Heatstroke

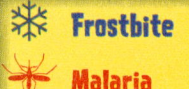

Frostbite and Hypothermia Bring layers of snug, water-resistant clothing, including warm mittens, wool or fleece hats, and sturdy water-resistant boots to regions with very cold temperatures. Stay warm, dry, and covered to prevent frostbite and hypothermia. Frostbite is the freezing of the skin and tissues under the skin. Often the skin will turn white or blue, and the tissue becomes damaged. This occurs at the extremities—hands, feet, and nose—when cold temperatures slow blood circulation. Hypothermia occurs when your internal body temperature drops lower than normal in an extremely cold environment. This is a dangerous condition in which all body functions slow down to the point of unconsciousness.

Malaria Transmitted by the bite of infected mosquitoes, malaria is a serious disease. If you are traveling to an area where the disease is present, have your doctor prescribe medication that prevents malaria. The first line of defense is to avoid mosquito bites. Sleep under a mosquito net and use insect repellent; also, wear long-sleeved shirts and tuck your pants into socks.

18 **DISCOVERY CHANNEL SCHOOL**

Tuberculosis A cough or sneeze from an infected person sends tuberculosis (TB) bacteria into the air. TB also spreads through milk from infected cows. The disease is a worldwide health threat, but Eastern Europe, southern Africa, and Southeast Asia have particularly high TB rates. Travelers visiting those areas should be tested when they return home. No vaccine is available to prevent infection, but prompt treatment can cure the disease.

Take the Safe Way

No matter where you are traveling, keep these tips in mind:
- Viruses, bacteria, and contaminated food or water often cause travelers' diarrhea. Before you go on a trip, take precautions. Read about the country you're going to, and consider drinking only bottled water or canned drinks, not tap water. Boil water before using it. Bring along some anti-diarrhea medication.
- Don't walk around barefoot and give ringworm and other parasites an opportunity to enter your bloodstream.
- Consult your doctor before you travel.

EUROPE

ASIA

Altitude Sickness The air at high elevations—8,000 feet (2,400 m) above sea level and higher—contains less oxygen. Spending time there may cause altitude sickness. The symptoms include shortness of breath, dizziness, fatigue, nausea, and impaired hearing, sight, and motor skills. Mountain climbers at extremely high altitudes of 20,000 feet or more carry extra oxygen to avoid the potentially fatal effects of altitude sickness.

AUSTRALIA

Activity

VITAL STATS The health dangers on these pages relate to climate and elevation, and they affect large areas of the globe. But other health concerns are specific to particular countries. Go to the Internet for the World Health Organization site www.who.int and make a list of the top three health issues for eight or ten nations. Be sure your list represents different continents. Which conditions are based on natural factors, such as climate? Which have origins in political or social issues?

Heatstroke If you are visiting a region with extremely high temperatures, you're at risk for heatstroke. This condition occurs when the body stops sweating and reaches too high a temperature. The result is dizziness, disorientation, and sometimes collapse. Protect yourself: Wear loose-fitting clothing, a cap or hat, sunglasses, and sunblock. Drink lots of bottled water to prevent dehydration and help the body retain water.

ANTARCTICA

THE HEALTH FILES

VIRTUAL VOYAGE

Liquor Lowdown

Alcohol often harms the body when consumed in large amounts, but the trouble begins on a small scale. How small? See for yourself, from an alcohol molecule's point of view.

You're a simple arrangement of carbon, hydrogen, and oxygen. And then someone lifts the glass in which you're swirling. With that person's swallow, you're suddenly slipping past teeth and tongue, down the esophagus to arrive in the stomach, or perhaps sliding farther down into the small intestine. The process is so simple it may be difficult to believe that a steady diet of too many molecules just like you can interfere with the absorption of important vitamins and minerals such as Vitamin A, Vitamin B12, magnesium, and thiamin.

Once in the bloodstream, it's not long before you're absorbed through the stomach. But you mostly travel to the small intestine's walls and into the bloodstream. You're then pumped through the heart and throughout the body. The presence of you and other alcohol molecules makes blood pressure rise. As few as two drinks a day can keep blood pressure high, enlarging the heart over time. In the long run, this process can damage the heart muscle, leading to heart disease and potentially heart failure.

Within 30 seconds, you've sped past the heart, through arteries, and on to the brain. Inside the brain is the cerebral cortex, which controls a person's use of reason and judgment. Once there, you make the person feel lightheaded and relaxed. The cerebral cortex nerves talk to each other by using impulses that travel through neurotransmitters. Your presence causes one of these neurotransmitters, gamma-aminobutyl acid (GABA), to shut down. Without this neurotransmitter, the person drinking might say and do things more impulsively than usual.

Dazed and Confused

You also act as a sedative, causing the drinker to feel relaxed. This false sense of well-being can hide serious problems. Little does the person know you're killing cells by the thousands. Your presence in the bloodstream fools the brain into believing the body is carrying too much water, and so it gets rid of more water than it should. You are also poison to various kinds of nerve cells, so brain cells die when they come in contact with you. You affect the brain's ability to

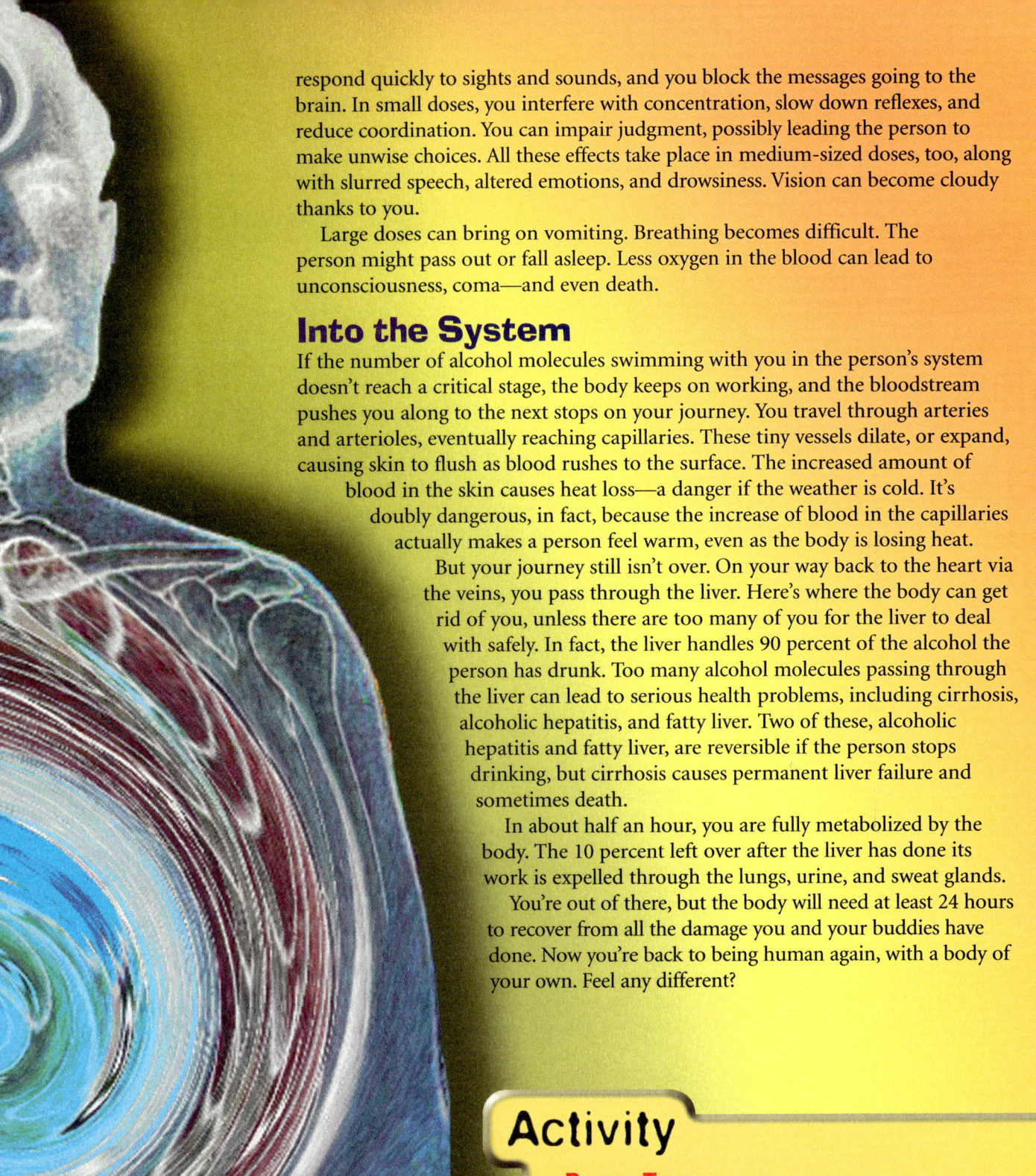

respond quickly to sights and sounds, and you block the messages going to the brain. In small doses, you interfere with concentration, slow down reflexes, and reduce coordination. You can impair judgment, possibly leading the person to make unwise choices. All these effects take place in medium-sized doses, too, along with slurred speech, altered emotions, and drowsiness. Vision can become cloudy thanks to you.

Large doses can bring on vomiting. Breathing becomes difficult. The person might pass out or fall asleep. Less oxygen in the blood can lead to unconsciousness, coma—and even death.

Into the System

If the number of alcohol molecules swimming with you in the person's system doesn't reach a critical stage, the body keeps on working, and the bloodstream pushes you along to the next stops on your journey. You travel through arteries and arterioles, eventually reaching capillaries. These tiny vessels dilate, or expand, causing skin to flush as blood rushes to the surface. The increased amount of blood in the skin causes heat loss—a danger if the weather is cold. It's doubly dangerous, in fact, because the increase of blood in the capillaries actually makes a person feel warm, even as the body is losing heat.

But your journey still isn't over. On your way back to the heart via the veins, you pass through the liver. Here's where the body can get rid of you, unless there are too many of you for the liver to deal with safely. In fact, the liver handles 90 percent of the alcohol the person has drunk. Too many alcohol molecules passing through the liver can lead to serious health problems, including cirrhosis, alcoholic hepatitis, and fatty liver. Two of these, alcoholic hepatitis and fatty liver, are reversible if the person stops drinking, but cirrhosis causes permanent liver failure and sometimes death.

In about half an hour, you are fully metabolized by the body. The 10 percent left over after the liver has done its work is expelled through the lungs, urine, and sweat glands.

You're out of there, but the body will need at least 24 hours to recover from all the damage you and your buddies have done. Now you're back to being human again, with a body of your own. Feel any different?

Activity

POSTER TIME Have a class competition to make an anti-drinking poster. Whoever makes the most visually interesting and scientifically accurate poster wins. The poster should contain some of the information that is provided on these pages. The purpose of the poster is to incorporate the science of alcohol with its dangers in the human body.

THE HEALTH FILES 21

SCIENTIST'S NOTEBOOK

BRAIN GAIN

New research reveals amazing things about the teenage brain.

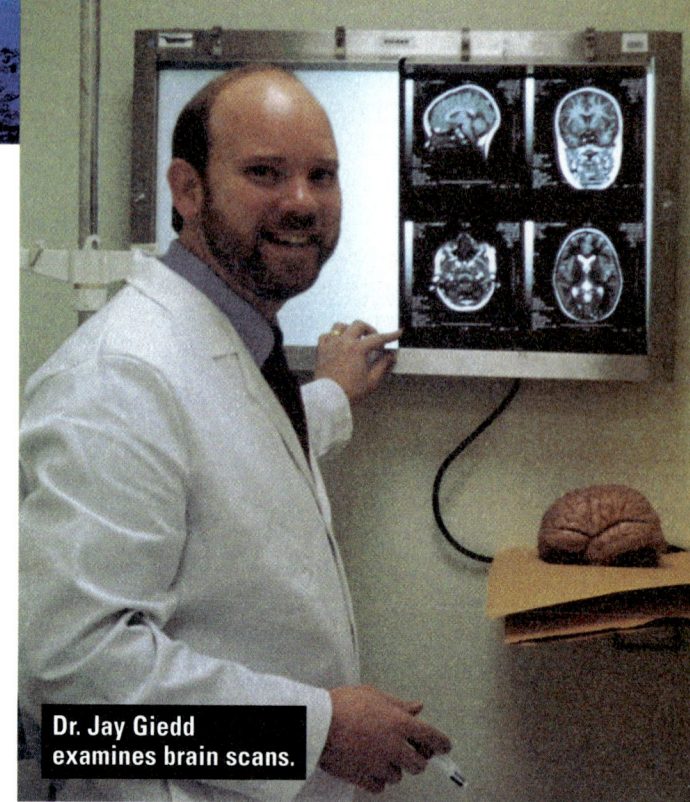

Dr. Jay Giedd examines brain scans.

Typical teen behavior is not always flattering. Teens flip-flop on decisions; their emotions are jumbled. Teens scream at their parents one minute, laugh with their friends the next. And for some teens, a risky ride in a friend's car sounds like a great way to spend an afternoon.

The cause of much of this chaos and confusion is located in the three pounds of tissue between the ears—the brain. Teen brains are not like the brains of adults and younger children. Dr. Jay Giedd is a psychiatrist who is mapping the differences.

Scientists once believed that the most important development of the brain took place before a baby was born and during its first few years. Researchers believed that a child's brain grew very little after the ages of five or six, and that areas of gray matter, where important brain-cell connections are made, began to decrease around that age. By the teen years about 100 billion neurons, or nerve cells, were performing the brain's different functions. The number of neurons, the size of the brain—these were set for life. Or so they thought.

In 1999 Giedd presented data that challenged these ideas. His discovery: Major areas of the brain continue to grow until about the age of 15. One significant spurt at the front part of the brain peaks at about age 11 for girls and 12 for boys. Other brain growth continues into the early 20s. Giedd believes the early teen years are crucial for kids, as their mental activities create connections in their brains that will last a lifetime. "The guiding principle," says Giedd, "is use it or lose it."

Snapshots of the Brain

Giedd works at the National Institute of Mental Health in Bethesda, Maryland, where he puts old theories about brain growth to the test using new technology. Magnetic Resonance Imaging (MRI) gives him a 3-D image of the brain.

Using MRI Giedd has studied the brains of almost 1,000 children since 1991. The process is harmless, though it requires the kids to keep still for about 10 minutes inside a large metal cylinder. Giedd has periodically scanned the brains of kids between 3 and 15 years old. Giedd's most recent images showed growth in gray matter at the corpus callosum part of the brain. This group of nerve fibers sends information between the brain's right and left hemispheres, or the cerebrum, which controls speech, hearing, and thought.

Grow For It

From ages three to six, the fastest growth takes place in the frontal lobe of the brain. This area controls the ability to plan and carry out new tasks and affects the control of emotions. As children grow older, the areas of growth move farther back into the brain. From age 6 to about 12, most growth takes place in the temporal and parietal lobes. These areas of the brain play a large part in language skills and spatial relations. Changes continue in the frontal lobe.

The brain always makes more gray matter than it needs. Any activity, such as reading or playing an instrument, creates connections between brain cells. The brain gets rid of cells that do not form connections, a process Giedd calls "pruning." During the teen years, gray matter in some very active areas of the brain can double. These are places where neurons are constantly and rapidly interconnecting, which may explain why teens are prone to wild mood swings or have trouble making decisions.

Giedd has found that changes also occur in the amygdala (ah MIG dah la), a small, primitive part of the brain that can direct our responses to fear and anger. During the teen years, the amygdala grows quickly in boys and girls, though more so in boys. It dominates the reaction to danger before other parts of the brain have had a chance to catch up, and before experience can teach lessons about weighing risks and benefits. Sometimes the instinctive response to danger or risk isn't the safest one. The fast growth of the amygdala in teens might help explain why teens take great risks and have a reduced sense of fear.

Good News for Teens

New information about brain growth is important to teens. As toddlers, children had no control over the activities that would help the growth of brain-cell connections. But teens get another chance to develop their brains. "Whether they do art, or music, or sports, or video games," says Giedd, "the brain is figuring out what it needs to survive and adapting accordingly."

Other scientists think the information on brain growth may affect education. The ages between 7 and 11 might be best for teaching kids to speak a foreign language or to play a musical instrument. But most scientists are still trying to determine which skills are best taught at a particular age.

In the meantime Giedd continues to scan young brains. "The current focus of our research is to determine what forces guide or influence [the] pruning process," he says. "Is it the activity of the teen, is it education, is it genetics, diet, medication?" Giedd keeps spreading the message he's already learned from his research: "The teenage years are a . . . critical time to optimize the brain."

Teens take more risks than adults.

Activity

BUILD A MODEL BRAIN To help understand the complexity of the human brain, you can create a model of it. You'll need a detailed diagram of the brain and some modeling clay of different colors. Start from the inside of the brain and build outward. Each distinct part should be a different color. Be sure to include the frontal lobe, parietal lobe, occipital lobe, brain stem, pituitary gland, and skull.
See http://faculty.washington.edu/chudler/lobe.html for more information.

AMAZING BUT TRUE

FOOD
for thought

The body is a complex machine. It knows just what to do with the food you eat.

Food gives you fuel, just as gas fuels a car engine. We measure energy from food in calories. These come in three tasty varieties: fats, proteins, and carbohydrates.

All cell membranes are made partially of fat, which also cushions us against falls and famines. If no food is coming in, the body uses back-up supplies of energy found in fat. Each gram of **fat** is packed with nine calories.

One gram of **protein** gives you four calories of energy, which your body uses to grow and maintain its functions. Protein strengthens your muscles, nails, and bones.

Carbohydrates are the fuel that keeps us moving. Everything you do—even thinking or sleeping—burns up some of the energy stored in your body, and carbohydrates are always the first to go. (You just used one calorie reading this.)

The "Vite" Stuff

All food, except maybe the cheese you spray out of a can, comes with small but important traces of vitamins and minerals. Without these we'd die. But it's hard to keep track of exact amounts of these nutrients. The best way to make sure you're covered nutrient-wise is to eat a wide variety of fresh fruits and vegetables, lean meat, low-fat dairy products, and lots of grains.

You'll lose out on nutrition if you eat too many processed foods. They've got the calories, plus a lot of other things you don't need, such as additives and preservatives. While these ingredients may add flavor and extend a food's shelf life, they don't generally contain nutrients. Processed foods may also have extra sodium (too much of which increases your blood pressure), sugars, and hydrogenated fats (fats loaded with extra hydrogen, which makes them harder for your body to digest).

Our bodies are pretty smart when it comes to directing us to what we need, but they can be confused by food that's been made to taste like what it's not. If you stick to food that hasn't strayed too far from its origins, your body will have an easier time sorting out what it needs and using the ingredients that are good for it. It's worth making that extra effort to give your body what it needs—not just what you "feel like" eating.

Foods with partially hydrogenated fats are very hard to digest and should be avoided.

Nutrition Facts
Serving Size: **3.5 cups (30g)**
Servings per bag: **3**
Calories: **170**
Fat Cal.: **100**

AMOUNT PER SERVING	% DV*
Total Fat 11g	17%
Saturated Fat 2g	10%
Cholesterol 0mg	0%
Sodium 180mg	8%
Total Carbohydrate 16g	5%
Dietary Fiber 3g	12%
Sugars	0g
Protein	2g
Vitamin A	0%
Vitamin C	0%
Calcium	0%
Iron	4%

*Percent Daily Values are based on a 2,000-calorie diet.

Too Much of a Good Thing?

How many calories do you need? The answer used to be "as many as you can get," but in 21st-century America, we need to know when to stop. In prehistoric times salt, sugar, and fat were rare delicacies. But now that all three are readily available, our stone-age instincts push us to overindulge.

Here's a general formula: lots of vitamins, not too much fat, a lot of carbs, a little protein, and plenty of water—eight 10-ounce glasses every day.

Carbohydrates from fruits, vegetables, and whole grains should make up about 60 percent of your daily calories. Go for food that hasn't been overly processed or refined. For example, whole grains like wheat bread and brown rice have more nutrients than the white versions. Try to think of the heavy white stuff—rice, pasta, potatoes, bread—as the center of your meal; that's what most of your calories should come from.

About 25 percent of your calories should come from fat, but because fat is more than twice as high in calories as protein or carbohydrates, it'll look like a lot smaller than a quarter of what's on your plate. Cooking food in oil or coating it lightly with butter is yummy, and it can make broccoli go down a lot easier.

The final 15 percent of calories in a well-balanced diet comes from protein. Foods high in protein are meat, fish, eggs, milk, and beans. Red meat, whole milk, and cheese are high in fat as well as protein, so avoid the all-cheeseburger-and-ice-cream-sundae diet. A great low-fat, high-protein food is fish.

Pie eating contests are not good for controlling calorie intake.

Activity

BALANCING ACT Look at the table below of recommended average weights for boys and girls ages 12–16. Why do you suppose there is such a range of average weights? Is your weight in the average range? If not, why do you suppose you are outside the average range? Are you taller than average? Shorter? What are your eating habits? At your next doctor visit, ask if you should be concerned about your weight.

BOYS		GIRLS	
AGE	WEIGHT	AGE	WEIGHT
12	70–112	12	72–123
14	90–145	14	90–145
16	112–172	16	100–158

THE HEALTH FILES   25

Mystery of the Misplaced Calories

Why was Graham gaining weight when he watched what he ate so carefully?

Graham Baker stepped off the scale and looked at himself in the mirror. Sure, he'd grown four inches in the past year, but he couldn't believe that he'd gained almost 30 pounds. This time last year he'd be at the pool swimming laps. But when his best friend dropped the swim team last March, Graham did, too. That was six months ago.

"I've got to do something," he said, feeling more self-conscious than ever in his swim trunks. "I might as well start now. It shouldn't be too hard to eat less and cut down on junk food. I'll keep a diary, too, so I can track my progress."

After following a new routine for about two months, Graham was surprised and disappointed by the results. He decided to pay a visit to the school nurse for some advice. He showed his food diary to Ms. Staywell.

"First of all," the nurse explained, "your body turns food into glucose for fuel. When your glucose levels drop, your body functions at a lower metabolic rate, which means you don't burn calories as quickly. Dropping glucose levels also contributes to sluggishness and lack of concentration.

"You've still got some growing to do," she went on. "Your body needs a lot of nutrients—something that dieting can deprive you of. But it's also very important to make healthy choices now, at a time when you can prevent diseases that might otherwise strike you 10, 30, even 50 years from now."

Ms. Staywell reviewed the diary. "You've been making some good choices about food," she said. "But I'll tell you why you're still gaining weight."

Review Graham's diary entries yourself and use the clues below to find out why he didn't lose any weight on his diet. Can you figure out what the nurse told him?

Graham's Food Diary

Day 1

I got off to a great start today by skipping breakfast. That gave me some extra time to get to school. It was a good thing, too—I needed time to study before my first-period math test! I thought I did okay, but during the second half of the test, my mind kept wandering. I ended up doing a lot of problems over several times trying to get the right answer.

Day 9

Instead of eating breakfast, I've been sleeping in a bit more. Seems like I need it, too. By lunchtime yesterday I was pretty down in the dumps. I guess because math class seems a lot harder this year. I was still hungry after one serving of macaroni and cheese. I figured since I already skipped a meal today, I could eat an extra helping, so I did. I also had a glass of milk and a fat-free brownie instead of my usual cola and candy bar. I feel like my diet has been going well, but I haven't lost any weight yet.

Day 32

Cal called after school to go bike riding today, but I just didn't feel like going. I guess I felt more like watching TV. Plus, I was hungry. I've still been good about avoiding junk foods, but I miss them! I really wanted some potato chips this afternoon, but I picked a nutritious snack—a bag of microwave popcorn. The whole bag only had 170 calories. That seemed pretty low. I still haven't lost any weight. In fact, I think I might have gained a pound.

Day 45

Today was a pretty good day. I'm determined to really cut back food to lose weight, because so far, my diet hasn't been working. I had a tuna fish sandwich and a salad for lunch today. Tonight we went to Aunt Suzy's for dinner and she made my favorite: spaghetti and meatballs with garlic bread! She uses fresh tomatoes in her sauce, so I didn't feel guilty about eating seconds.

Day 63

I can't believe it. I've been so careful about eating healthy and I've gained three pounds. I don't get it. Last year I ate so much more and I felt okay in my swim trunks, too. Why is it so hard to lose weight?

Clues

Use these clues:

1. **Lunch, Day 9**
   - 2 helpings of macaroni and cheese (975 cal.)
   - 1 cup steamed green beans with butter (125 cal.)
   - 8 oz. whole milk (210 cal.)
   - fat-free brownie (120 cal.)

2. **Snack, Day 32**
   - 1 bag of microwave popcorn
   - Servings per bag: 3
   - Calories per serving: 170

3. For his height and build, Graham needs about 2,000 calories every day.

4. The body's level of activity affects its need for calories. An active teen may need 1,800 calories, but a less active teen may need fewer than 1,800 calories.

Answer on page 32

THE HEALTH FILES **27**

D.A.R.E. to Be Different

The Power Behind Drug Awareness Resistance Education

> This does not just happen in ghettos or inner cities. It does not just happen to poor kids or kids from broken homes. It happens everywhere in America . . . to kids of every age, gender, race, and social class.
> —Glenn Levant
> President and founding director of D.A.R.E. America Worldwide

"Hey! Want a hit?"
"Just try it!"

Kids get this offer on a daily basis in many schools throughout the United States. The answer they give could change their lives forever. Thanks to a school program called D.A.R.E. (Drug Awareness Resistance Education), kids are learning how to respond "no."

Glenn Levant started D.A.R.E. in Los Angeles in September 1983. Undercover operations that put officers in schools to buy drugs and make arrests made little progress in eliminating drug use. D.A.R.E. hit the scene with a radical new approach: Get to kids before they start using drugs.

"I was convinced . . . it is the only solution—eliminate demand. You can't sell something no one wants to buy," said Levant, president and founding director of D.A.R.E. America Worldwide.

D.A.R.E. police officers work with fifth-grade students.

Cops in the Classroom

The D.A.R.E. program works by teaching students practical ways to resist drugs. Police officers bring the anti-drug message to classrooms. "If you put a streetwise, specially trained police officer into the classroom . . . kids will pay attention," said Levant. D.A.R.E. is a partnership between the Los Angeles Police Department and the Los Angeles Unified School District.

Police officers bring to the classroom their day-to-day experience with drugs and violence. According to Levant, "Officers . . . can answer a fifth-grader's question about the difference between 'crank' and 'crack' and explain what it does to a young mind and body."

Each participating school has its own D.A.R.E. officer, whom students get to know and trust. Students know they can talk to the officer as a sympathetic counselor. "We're building a rapport with them," said Detective Thomas Hegel, who teaches D.A.R.E. classes.

"Middle school is usually when kids start with cigarette smoking . . . and gateway drugs," he says. D.A.R.E. pays special attention to fifth- and sixth-graders. Hegel explains why it's important to educate kids before they start drinking, smoking, or experimenting with drugs. Kids this age are pressured to fit in, and they are easily influenced by peers and older siblings. Television shows, movies, and music sometimes make drugs and violence look cool.

Pressures and Temptations

Most kids who try drugs are afraid of saying "no." D.A.R.E officers help kids act out typical drug-exchange scenes. The exercises teach kids how to turn down drugs: Give a reason or excuse ("I don't smoke"), walk away while saying no, change the subject (suggest another activity), or avoid the situation (stay away from places where kids are using drugs).

Older students act as mentors, showing that drugs and violence are not a part of being accepted. "Many of [the students] think that high school means getting into fights," Hegel explains. "The peer leaders tell them that it's not so, that there are alternatives."

Officers in the program recognize that kids face many pressures with school, athletics, difficult issues with their friends, or family problems. Drug use and violence can result when teens don't know how to handle these pressures. The D.A.R.E. officers give students someone to talk to about handling problems. "Sometimes we play kickball in the yard to show that we can have fun without drugs," Hegel says.

Teens march with D.A.R.E. officers in a parade.

Looking Forward

Almost two decades after it began, D.A.R.E. has had a positive effect on kids' attitudes about drugs. Drug use by teens is on a steady downward trend. Almost 90 percent of D.A.R.E. students have reported that the program gave them skills and self-confidence to avoid drugs and alcohol.

Today the program is in 80 percent of the school districts in the U. S. and in 51 other countries as well. In a typical year D.A.R.E. will encourage 36 million kids to choose not to "follow the crowd." The war on drugs isn't over yet, but D.A.R.E. and its dedicated officers are winning heroic battles in classrooms every day, all over the world.

Activity

D.A.R.E. DRAMA Create an anti-drug play that is appropriate for grades 4 and 5. Write a script, get scenery, and pick the actors. You have to decide what kind of story would be appropriate for this age group, and what the play will be about. Present your play to your teacher and work on any modifications. When you have gotten all the proper permission, perform this anti-drug play for kids in your school.

THE HEALTH FILES

HEALTHY FUN

FUN & FANTASTIC

Don't play games with your health. But you can play this game for a healthy dose of fun. Grab some friends, a number cube or die, and some common objects (e.g., coins, buttons, erasers) for game pieces. Get rolling!

START

Directions
The object of the game is to be the first one to finish. Roll the die or flip a coin to decide the order of players. Then each person rolls the die and advances on the game board; the landing square indicates the next move.

Drank beer with friends and got sick. Go back three spaces.

Cut gym class and smoked cigarettes with a friend. Go back five spaces.

Warned your parents or a teacher about a friend who might have an eating disorder. Go forward three spaces.

Ran across the street against a red light. Lose a turn.

Helped make burger patties at the family barbecue but didn't wash your hands. Start over.

At lunch with your friends, chose mixed salad instead of French fries. Go again.

Took a shortcut on your bike—the wrong way against one-way traffic. Lose a turn.

30 DISCOVERY CHANNEL SCHOOL

YOUR WORLD YOUR TURN

Safe and Sound

In California lead poisoning is the most common chemical hazard for kids. As of 1992 about 239,000 kids in that state were found to have levels of lead in their blood high enough to put them at a health risk. Lead exists in many everyday materials: paint, soil, food, air, plumbing fixtures, water, and even some hobby products. To protect kids from lead hazards and to spread the word about lead-poisoning prevention, the Childhood Lead Poisoning Prevention Program went into effect that same year. Since then, the public has become aware of lead-based hazards, and kids and families have taken an active role in making sure their environment is free of lead.

Lead isn't the only chemical hazard. Asbestos, a building insulation material, and radon gas are also major concerns in schools and older homes. Knowing about these hazards—and what you can do to prevent exposure—is important.

How safe is your school, your home, or your own backyard? These are the places where you hang out most of the time—and they should be safe places for you and your friends. But there may be some hidden dangers you don't know about. Track down potential dangers by becoming a health detective. Here's what to do:

1. Use the library and Internet to learn more about lead, asbestos, radon gas, and other potentially dangerous substances in the environment. Make a checklist containing everything you want to look for in your home.
2. Talk to your school principal or custodian to learn the guidelines and processes used in your school system to monitor environmental and chemical hazards. Ask to get a copy of the last report written about your school.
3. To get information about your home, ask a parent to discuss your checklist and explain if any of the items apply to your house.
4. Contact the public safety department in your city or county. Ask for the checklist inspectors use when they examine buildings. Compare this to your checklist.
5. Suggest ways you and your family can help improve the health and safety standards of your home. Small steps can make big results.

Ready for the ultimate challenge? Enter this or any other science project in the Discovery Young Scientist Challenge. Visit http://school.discovery.com/sciencefaircentral/dysc/index.html to find out how.

ANSWER Solve-It-Yourself Mystery, pages 26–27

Ms. Staywell told Graham that his first mistake was skipping breakfast. His energy levels continued to drop in the mornings, along with his metabolism. When Graham's brain didn't get new fuel in the morning, hormones told his brain that he needed more calories. That made it hard for him to concentrate during math class.

Skipping morning meals contributed to Graham's habit of overeating later in the day. His macaroni and cheese lunch contained 1,430 calories—nearly 75 percent of the total number of calories he needs in a single day. He would have to eat very little for the rest of the day to avoid taking in more calories than he should in a single day.

Graham thought he made a better choice with popcorn over potato chips. But he misread the label. The number of calories—170—is based on one serving, and the bag contains three servings. Graham ate the whole bag so he actually consumed 510 calories. With habits like these, Graham could have easily eaten an extra 167 calories a day—enough to gain three pounds in 63 days!

Graham's best bet would be to get back on the swim team and eat regular, balanced meals. By continuing to choose more healthful snacks over junk food, he is more likely to get all the nutrients he needs to do better in math as well as win a ribbon at the swim meet (and feel great in his swim trunks, too).